Exoplanets

by

Patrick H. Stakem

(c) 2018

Number 23 in the Space Series

Table of Contents

Introduction..4
Author...6
Introduction..7
The diversity in our own solar system................................7
    Mercury...8
    Venus..9
    Earth..9
    Mars...10
    The Asteroid Belt...11
    Jupiter...11
    Saturn...12
    Uranus..12
    Neptune..13
    Pluto and beyond..14
    Dwarf Planets...15
The Drake Equation...15
ExoPlanets...17
    Ross-128b..18
    Trappist-1...19
    Beta Pictoris b...19
    Gilese 581..19
    H209458b...20
    WASP-3b..20
    PSR B1620-26..20
    OTS-44..20
    Barnards Star-B..21
    Alpha Centauri's planets..21
    Kepler-443b...21
Search for Exoplanets...21
    Earth based 24
    Space-based...25
        HST..26
        Wfirst...28
        Kepler 28

- NASA's TESS Mission..30
- Spitzer...30
- JWST..31
- HabEx...33
- Cheops..34
- ESA's PLATO Mission...34
- COROT...35
- ESA's GAIA Mission..35
- European Ariel Mission..35
- Canada's MOST Satellite..35
- Europa Clipper Mission..36

Interstellar probe...36
Afterword..37
Glossary..38
Bibliography...44
Resources..48
If you enjoyed this book, you might also be interested in some of these..50

"A single ear of wheat in a large field is as strange as a single world in infinite space."

Attributed to the 4th Century BC Skeptic, Metrodorus of Chios, who had an amazing cosmic view.

"if the fixed stars are the centres of similar systems, they will all be constructed according to a similar design and subject to the dominion of One. Isaac Newton, conclusion of Philosophiae Naturalis Principia Mathematica.

Introduction

This book covers the topics of the existence of planets around others stars. We think now we are unique in the Universe, but the Universe is very large. This book just touches the surface of this very exciting field. With better telescopes and data processing methods, more planets in our galaxy have been found. Dedicated planet-finding facilities on the ground and in space have been deployed. We now see that planets are rather common in our galaxy. The next big question is, do any of them have life?

One of the more exciting missions is the search for planets of other stars than our own sun. Although there are nearly impossible to image directly, they can be observed as they pass through our line of sight with the distant star, and cause a small dip in the perceived brightness. Exoplanets are best seen from space.

Some evidence for exoplanets was found as early as 1917, but detection of one didn't happen until 1988.

In the year 2,000 the discovery of exoplanets began for real, as new technology was deployed. Older programs, such as SETI (Search for Extra-Terrestial Intelligence) had been searching the radio spectrum for some time. There are more un-inhabited planets that we know of (3,874) than those with radio transmitters (1). exoplanets come in all sizes, and about 1 in 5 stars like our Sun have one similar to Earth. Do the math. With 200 billion stars in the Milky Way along, that could mean potentially 11 billion potential Earth's out there. On the average, each star has one planet. In addition, about 1 in 5 stars like our Sun, have a planet in the "habitable zone." In a binary star system, an exo-planet can orbit either or both stars. There are exoplanets in triple star systems, and at least one has been observed to have a ring system like Jupiter and Saturn. Exo-moons orbit exoplanets. Atmospheres have been discovered around some exoplanets. Some exoplanets are tidally locked to their primary, much like Mercury in our system. Exoplanets can be "rocky" like Venus, Earth, and Mars, or "gassy," like Jupiter and Saturn.

A few thousand have been cataloged. Can exoplanets harbor life? To have the potential of life, the planets have to orbit the right type of star, at the right distance, called the Goldilocks zone (from the Fairy tale). Not too hot, not too cold, just right.

Exoplanets are the current hot topic, and dedicated missions have been launched to search for and characterize them. The number of interesting known exoplanets is literally changing on a daily basis. When the new James Webb Space Telescope gets launched in a few years, the floodgates will open wider.

In 1980, there was a 12-year program kicked off to find exoplanets around regular stars, with no results. The first exo-planet was discovered in 1992. It's host star was a pulsar, with two planets. Irregularities in normally regular pulsars was the key to the discovery. Of these, 2893 are transiting their host star, from our point of view. Most exoplanets discovered are around main sequence stars. The first exoplanet found around a regular star like our Sun was seen around a star in the constellation Pegasus.

The Hertzsprung–Russell diagram is a graphical representation of stars plotted by luminosity versus temperature. It was created in 1910. It shows clustering in 4 areas, the main sequence of stars, white dwarfs, giants, and supergiants. There are outliers, but most stars follow the clustering.

As of this writing, there are more than 3,800 known exoplanets, with more being added every day. There are 600 known multi-planet systems. I will update this book periodically to include recent discoveries.

At the moment, it is easy to get money to search for exoplanets. Congress has mandated NASA to do just that. They are to: "acquire an improved understanding of how planetary systems form and evolve, including better descriptions of planetary system architectures, compositions, and environments. Second, they need to learn enough about exoplanets to make informed predictions about habitability, and to make meaningful searches of alien life in distant star systems. "

Suggested focus areas include a new generation of both Earth-based and space telescopes. They do realize that they can't tax exoplanets, right?

Author

The author has a BSEE in Electrical Engineering from Carnegie-Mellon University, and Masters Degrees in Applied Physics and Computer Science from the Johns Hopkins University. During a career as a NASA support contractor from 1971 to 2013, he worked at all of the NASA Centers. He served as a mentor for the NASA/GSFC Summer Robotics Engineering Boot Camp at GSFC for 2 years. He teaches Embedded Systems for the Johns Hopkins University, Engineering for Professionals Program, and has done several summer Cubesat Programs at the undergraduate and graduate level.

He began his career in Aerospace with Fairchild Industries on the ATS-6 (Applications Technology Satellite-6), program, a communication satellite that developed much of the technology for the TDRSS (Tracking and Data Relay Satellite System). At Fairchild, Mr. Stakem made the amazing discovery that computers were put onboard the spacecraft. He quickly made himself the expert on their support. He followed the ATS-6 Program through its operation phase, and worked on other projects at NASA's

Goddard Space Flight Center including the Hubble Space Telescope, the International Ultraviolet Explorer (IUE), the Solar Maximum Mission (SMM), some of the Landsat missions, and others. He was posted to NASA's Jet Propulsion Laboratory for the MARS-Jupiter-Saturn (MJS-77), which later became the Voyager mission, which is still operating and returning data from outside the solar system at this writing.

He received NASA's Space Shuttle Program Managers Commendation award, two NASA Group Achievement Awards, and the NASA Apollo-Soyuz Test Program Award. He has completed over 42 NASA Certification Courses. He has led and supported efforts at all of the NASA Centers on terrestrial and planetary missions.

Mr. Stakem has been affiliated with the Whiting School of Engineering of the Johns Hopkins University since 2007. He supported the Summer Engineering Bootcamp Projects at Goddard Space Flight Center for 2 years.

Introduction

The International Astronomical Union defined in August 2006 that a planet is a celestial body which is in orbit around a star, has sufficient mass to assume hydro-static equilibrium (i.e. a nearly round shape) and. has cleared the debris from around its orbit. In addition, the IAU now controls the naming process for exo-planets. For the case of exoplanets orbiting a single star, the planets used the star name followed by a letter. The brightest planet gets the letter A.There is a provisional agreement on naming exo-planets orbiting multiple stars. The first planet discovered gets "B".The process continues through the alphabetwith the next to be discovered getting the next letter. If several are discovered at the same time, the closest gets the next letter. There is a procedure for exoplanets orbiting one star of a binary star system, and theu're working on the protocol for exoplanets orbiting more than one star. In addition, there is a protocol in place for actually naming the exoplanets.

The diversity in our own solar system

Before we begin looking at Exoplanets, let's look at the planets in our solar system, to get a baseline of what we're looking for. The four innermost planets are rather small, and rocky. Only one is known to host life. The four outer planets are large and made of ice or gas. All but two are known to have moons. The Sun is a star, about which the planets in our Solar Systemrotate. It provides energy to the planets. It is a stable G-type star in the Main Sequence.

All the planets in our Solar System orbit the Sun in a counterclockwise direction as viewed from above Earth's north pole. Most planets also rotate on their axes in an anti-clockwise direction, but Venus rotates clockwise.

Mercury.

Mercury, the closest planet to the Sun, is in tidal lock, with one side always facing the Sun. Actually, there is a rare 3:2 spin-orbit resonance, not seen elsewhere. For every two revolutions around the Sun, Mercury rotates three times on its axis. It wobbles a bit, creating a twilight zone that is much less extreme. It has no known moons, or Trojans. Being so close to the Sun, it is difficult to observe the planet and its immediate vicinity. It has a heavily cratered surface. Mercury is a rocky planet, with long narrow ridges observed on the surface.

Mercury has a molten core with abundant iron. Its mantle is made of silicates. The crust is thought to be some 35 km thick.

The surface of the planet resembles Earth's moon, with large plains (mare) and craters. It is thought to be 4.6 billion years old. It has no atmosphere.

Mercury is currently being observed up close by the Messenger spacecraft, and this will increase our knowledge of the planet. We really don't have a way to image or study the sun-facing side. The spacecraft has seen evidence of more than 50 pyroclastic flows from active volcanoes.

The surface temperature of Mercury varies from -170 to 435 degrees Celsius. Water ice has been confirmed in deep craters at the poles. Mercury has no atmosphere per se, due to its low gravity. It does have a magentosphere, with a field strength about 1% of Earth's.

The eccentricity of Mercury's orbit is 0.2, the most of any planet in our solar system. The axial tilt is almost zero. Mercury's orbit is tilted 7 degrees to the plane of the ecliptic. The orbital eccentricity varies wildly from zero to 0.45 over a period of millions of years.

Venus

Heavy greenhouse clouds of sulfuric acid trap solar energy, and cause massive global warming on a planetary scale. The surface temperature is high enough to melt some metals. We need to find out what went wrong on Venus, and try to avoid that on Earth.

Venus' atmosphere is 96% carbon dioxide at a surface pressure of nearly 100 times Earth's, a greenhouse gone wild. It has no moons. Venus is roughly Earth-sized, but something went terribly wrong

The heavily clouded atmosphere makes it difficult to observe Venus. We do know it rotates in the opposite direction to most of the other planets. It has no magnetic field.

Venus is a terrestrial, rocky planet, about the same size as Earth. It has a dense atmosphere of carbon dioxide. It's surface pressure is more than 90 times that of Earth's. Venus is hotter than Mercury in spite of being further from the Sun. It has seen extreme volcanism, but no lave flows have been observed. The surface was shaped by volcanic activity, and Venus has more volcanos than Earth.

Not much is know about the internal structure of Venus. There have been observations of lightning in the thick cloud layer. It has very high wind speeds.

Venus orbits the Sun in about 225 (Earth) days, with a small eccentricity. The planet has no moons. Due to extreme conditions, we can rule out life on Venus, as we know it..

Earth

We observe the Earth continuously with a series of orbiting satellites, to keep track of the weather, and violent events. In fact, Earth is the best observed planet, and we know the most about it. Not that it can't give us surprises. Earth's moon is the largest object in the sky, and has been observed since humans looked up. It is orbited by a series of satellites, has surface landers, and has been visited by Astronauts from the United States.

The bow shock is plasma from the Sun hitting the Earth's or other planet's magnetosphere. The plasma is ionized, and follows spiral paths along magnetic field lines. The flow speed, at Earth, is around 400 km/s. The shock, at Earth, is some 17 km thick, and located 90,000 km sunward. Bow shocks exist on planets with a magnetic field, and have been observed in other star systems.

Based on one data point, the only life forms that we are aware of are...us. And the animals and plants. It we see an exo-planet of roughly the size of the Earth, orbiting a star much like our Sun, at a comparable distance, we might conjecture the other planet might harbor life, like us.

Mars

Mars, and its two tiny moons and seven Trojans has got some infrastructure in place – a communications relay and a weather satellite. There are several Rovers and landers on the surface. It will probably be the first planet beyond Earth, to be visited, if not colonized, by humans

The Viking program was a pair of spacecraft sent to Mars in 1975. Each spacecraft consisted of an orbiter, and a lander. A major target now is a Mars sample return mission.

Mars is the second smallest planet, Mercury being the smallest. It is called the Red planet due to its color, caused by an abundance of iron oxide. It is a rocky planet, like Earth. It has a very thin atmosphere, and surface water is scarce, if present at all. There is sub-surface water ice, and ice at the poles. It's rotation period is Earth-like, and it has seasons. It hosts the largest volcano in the solar system. It has two moons, and a Trojan. The surface air pressure is about 1% of what we're used to. Because of the thin atmosphere, the temperatures vary wildly and there are planet-wide dust storms.

The surface gravity is about 38% of Earth's. It has a dense core, but no magnetic field. The dust on the surface is mostly iron oxide, but traces of other elements as well. There are impact craters, as its thin atmosphere is useless in burning up meteors.

Exploration of Mars is ongoing in-situ, and crewed missions are in the planning stage. Two big questions are: did Mars ever host life, and, can we live there?

The Asteroid Belt

There are millions of asteroids, mostly in the inner solar system. The main asteroid belt is between Mars and Jupiter. Each asteroid may be unique, and some may provide needed raw materials for Earth's use. There are three main classifications: carbon-rich, stony, and metallic.

The physical composition of asteroids is varied and poorly understood. Ceres, the Dwarf planet in the Asteroid belt, appears to be composed of a rocky core covered by an icy mantle, whereas Vesta may have a nickel-iron core. Hygiea appears to have a uniformly primitive composition of carbonaceous chondrite. Somewhat smaller than Ceres are Vesta and Passas. Many of the smaller asteroids are piles of rubble held together loosely by gravity. Some have moons themselves, or are co-orbiting binary asteroids. The asteroids rotate in their orbits, much like the planets.

The bottom line is, asteroids are numerous and diverse. We would expect to see similar groupings in other star systems.

The asteroids are not uniformly distributed. In the asteroid belt, the Kirkwood gaps are relatively empty spots. This is caused by orbital resonance of the asteroids with Jupiter.

Jupiter

Jupiter, the largest planet in our solar system, has 79 known moons, and perhaps 1 million Trojans of 1 kilometer or larger. The moons are diverse, and some may be water worlds. These tend to congregate at the L4 and L5 Lagrange points. The largest has a diameter of several hundred kilometers. The International Astronomical Union announced as this book was being prepared the discovery of 12 previously unknown moons of Jupiter, by an observatory high in the Andes in Chile. Only one has been named so far, after Valetudo, a great-granddaughter of Jupiter.

Jupiter and its neighbor Saturn are gas giants. This means they are mostly hydrogen and helium. They were close to becoming stars, but just didn't quite make the cut. There are five classes of gas giants known, based on their physical properties, ammonia clouds, water clouds, cloudless , alkali-metal clouds, and silicate clouds. Jupiter and Saturn both have ammonia clouds. The class of exo-planets know as hot Jupitershave the alkali metal or silicate clouds.

Jupiter probably has a rocky core, surrounded by liquid metallic hydrogen.

Saturn

Saturn and it's 62 known moons have been visited by spacecraft four times. The first was a flyby by Pioneer-10 in 1979. This showed the temperature of the planet was 250 degrees K. Voyager-1 visited in 1980. It conducted a close flyby of the moon Titan to study its atmosphere. It is, unfortunately, opaque in visible light.

We do know it rains methane. Voyager-2 swung by a year later, and data showed changes in the rings since its sister mission visited the year before. Temperature and pressure profiles of the atmosphere were gathered. Saturn's temperature was measured at 70 degrees above absolute zero at the top of the clouds, and -130 C near the surface. The flybys discovered additional moons, and small gaps in the rings.

Uranus

Uranus has 27 known moons, a 13-ring system, and a one-way light time of 2.7 hours from Earth. It has one known Trojan. Uranus was imaged in a flyby by the Voyager-2 spacecraft in 1986. It also captured some images of the Uranian moon Umbriel. But, Uranus and Neptune are two of the great remaining unknowns in our solar system, since neither have been explored in detail, by a dedicated mission. There is a desire to put an explorer spacecraft in orbit, and use that as a platform to launch probes into the atmosphere.

Uranus and Neptune are sometimes referred to as ice giants, since their atmospheres are known to contain water, ammonia, and methane ice. Uranus has a magnetic field. Interestingly, Uranus' spin axis is tilted into the plane of its orbit around the Sun. Seasonal changes and weather have been observed. The Voyager-2 mission imaged Uranus on its way from Jupiter, and out of the solar system. Atmospheric wind speeds are know to approach 900 kilometers per hour. It's orbit period is 84 Earth years. It receives about $1/400^{th}$ of the light that the Earth does from the Sun, so solar power is probably not a viable choice.

Because of the strange orientation of the planet's rotation axis, during the solstice, one side of the planet faces the Sun continuously, and the other faces deep space. Each pole gets 42 years of direct (though weak) sunlight, and 42 years of darkness. In spite of this, the equator is the hottest region. At this writing, the planet is in its autumnal equinox.

Uranus has a strange predominately water-ammonia ocean, which is electrically conductive. A major targeted mission is the Uranus orbiter and Probe. Mission analysis comes up with a 12-13 year cruise from Earth to Uranus.

Neptune

Neptune has 14 known moons, and 18 known Trojans. It was also visited by Voyager-2 in 1989. Six new moons were discovered. That is the extent of close-up observations of the planet. Neptune has rings, like Jupiter and Saturn, and a great dark spot. It's moon Triton has geysers and polar caps. Triton has an interesting retrograde orbit – it goes in a different direction than the other moons. Triton's surface is mostly frozen nitrogen, and is geologically active. It is speculated that Triton has a subterranean ocean. The moon Ptoteus is an ellipsoid, not a sphere.

Pluto and beyond

Pluto was downgraded from a planet to a Kuiper Belt object. The *New Horizons* mission to Pluto and the Kuiper Belt began in January of 2006, and reached the vicinity of Pluto in July 2015. It conducted a 6-month survey of Pluto, and went out farther into the Kuiper belt, on an 3 year extended mission, which is ongoing at this writing. The spacecraft was developed for NASA by the Johns Hopkins University Applied Physics Lab in Laurel, MD.

Pluto had one known moon, Charon, before the New Horizons Team members, using Hubble Space Telescope data, discovered four more, Nix, Hydra, Styx, and Kerebos.

There is lots of interesting stuff beyond Pluto, before Inter-stellar space is reached. The minor planet 90377 Sedna is located at about 86 AU and is one of the Trans-Neptunium Objects.

The Scattered Disk is a group of objects in elongated orbits, caused by the influence of Neptune. It extends from 30 AU to about 100

AU.

The Kuiper belt is a circumstellar disk, extending from Neptune (30 AU) to 500 AU. It resembles the asteroid belt. It holds mostly small bodies.

The Oort cloud is a circumstellar disc around our Sun, containing icy worlds. It extends from 2,000 to 200,000 AU. There is a disk-shaped inner cloud, and a spherical outer cloud. These, strictly speaking, do not belong to our solar system, but are located in interstellar space. It is the domain of comets.

Dwarf Planets

The dwarf planets of our solar system include Ceres. Orcus, Pluto, Salacia, Varuna, Haumea, Quaoar, Makemake, 2007 OR10, Eris, and Sedna. These smaller objects did not make the size cut to be a real planet. These all orbit the Sun. Ceres is located in the asteroid belt. Orcus is a trans-Neptunian object, Salacia, Haumea, Quaoar, Makemake, and Varuna are Kuiper Belt objects. Eris is the largest of the dwarf planets, having its own moon. Sedna is beyond the Kuiper belt. It's orbital period "year: is 11,400 Earth years. It's in a highly elongated orbit, probably due to Neptune's gravity. Generally, a dwarf planet does not have enough gravity to clear its orbit of other material. Not all dwarf planets have yet been discovered or observed. There may be 10's of thousands.

It is a reasonable assumption that other solar systems are organized similarly to ours. At least, that's a good first guess.

The Drake Equation

The Drake equation, formulated by Dr. Frank Drake in 1961, is a way of estimating the number of extraterrestrial Civilizations that can communicate. It was formulated by Dr. Drake at the Greenbank Radio observatory. The equation multiplies these factors:

R*, the average rate of star formation in our galaxy.

fp, the fraction of those stars that have planets.

ne, the average number of planets that can potentially support life per star that has planets.

fl, the fraction of planets that could support life that actually develop life at some point.

fi, the fraction of planets with life that actually go on to develop intelligent life.

fc, the fraction of civilizations that develop a technology that releases detectable signs of their existence into space.

L, the length of time for which such civilizations release detectable signals into space.

N, the number of civilizations in our Galaxy that ca communicate.

In 1961, guessing at the parameters, the value for N was 10^3-10^8. As more data is collected, better values for the parameters evolve. The current value is around 15 million.

The resultant figure for the universe suggests it is highly unlikely that Earth hosts the only intelligent life that has ever occurred. Maybe we can check with our neighbors...

The Fermi paradox, named after physicist Enrico Fermi, is the apparent contradiction between the lack of evidence and high probability estimates for the existence of extraterrestrial civilization. What are we missing?

Initially, we had to guess at most of these parameters, but better observation has brought about better results. As our observation technology gets better, we get better numbers for the factors in the equation.

In 1992, the first discovery that our solar system was not unique in the Galaxy was made. This gave us some real numbers to plug into the Drake equation. This observation was done by radio-telescope.

The Red Dwarf star Trappist has a system of 7 planets, nearly all

Earth size. That a red dwarf would have a planetary system was a surprise, but opened up the options for further research. At the moment, we can only detect exoplanets of Earth-size or larger.

Dr. Rene Doyon of the University of Montreal, with an International Team, photoed three XO planets orbiting the same star. This is the first detected X0 solar system. He used the Keck and Gemini North telescopes in Hawaii, on the summit of the volcano Mauna Kea. Both of these have 10 meter primary mirrors.

ExoPlanets

ExoPlanets are planets of other stars. Although it is difficult to see them through a telescope, we can define them by their gravitational effects on their primary (star).

A Galaxy, like our own Milky Way, is a collection of stars, gas, dust, and dark matter, gravitationaly bound together. The objects orbit the galaxy's center of mass, in an elliptical, spiral, or irregular shape. Between the stars and other objects in a galaxy is a gas, with a density around 1 atom per cubic meter. There are estimated to be around 10^{11} - 10^{12} galaxy's in our observable universe, each having 10^8 to 10^{14} stars.

At the moment, we know of around 4,00 planets orbiting other stars. We don't know how many are in the habitable zone. Even the definition of "habitable" is not well defined, and goes according to "life as we know it." Exoplanets can be in any orbit – they can rotate in the same direction they are moving in orbit, or the opposite way (retrograde). There are also rogue planets, that do not orbit a star, but rather the galactic center. It is thought that there may be a billion of these in the Milky Way galaxy alone.

We are just beginning to understand planet formation. We think we understand star formation.

More than one exo-planet can orbit a star, giving us exo-solar systems. In addition, an exo-moon has been discovered orbiting a known exoplanet. Recently discovered was something that had never been seen before, and needed a name. It was a small body, orbiting the moon of a known exoplanet. The astronomers decided to call it a *moonmoon*. Beyond that, seven known exoplanets orbit a binary star system. There can be two planets that orbit each other, without a star.

Some observed exoplanets are rocky, like Earth and the inner planets. Some as gas giants, like Jupiter and Saturn. Some are icy giants, like Neptune and Uranus.

There is an upper limit to the size of a planet. If it is more than about 12 Jupiter masses, it will evolve into a star, not a planet. The gravity at that level is enough to crush matter, and start nuclear fusion. This rather small star is called a brown dwarf.

There are many planet classifications, including Giant, Super Earth, Super-Jupiter, Sub-Earth, Mini-Neptune, a Planetar (brown dwarf or smaller), Planemo (not quite a star), and Mesoplanet, smaller than Mercury, but larger than Ceres.

There are so many exoplanets known that I will only discuss a few. The exoplanet takes on the name of its parent star, with a number representing the order in which it was discovered.

The largest exo-planet observed is HR2562, with a mass about 30 times that of Jupiter, On the other side of the spectrum, the smallest observed is about twice the size of Earth's moon. The nearest, discussed below, orbits our neighbor star Proxima Centauri, a bit more than 4 light years away.

This section will discuss a few representative exoplanets.

Ross-128b

Ross-128b is a confirmed exoplanet in the habitable zone of the star Ross-128. That star is about 11 light-years away. The planet was discovered in decades of data from the European Southern Observatory in Chile. It is thought that the planet is a rocky world. It does not transit its "sun," so no-traditional methods were used to surmise its existence. Also, no current telescope can discern its existence. Astronomers were able to discern the presence and abundances of several chemicals, from spectroscopy. These included iron, carbon, magnesium, aluminum, calcium, potassium, and titanium. It's host star is smaller than our sun, but the planet orbits closer, so it falls in a temperate zone, possibly conducive for life. It is one of the most Earth-like exoplanets found to date.

Trappist-1

Trappist-1 is a ultra-cool red dwarf star slightly larger than Jupiter. It's almost 40 light years away. It has a nice solar system of exoplanets, with seven planets, three of them in the temperate zone, the record to date.

Beta Pictoris b

Beta Pictoris b is a known exoplanet orbiting the star Beta Pictoris in the southern hemisphere constellation Pictor. The star is some 63 light-years from Earth. The planet is a super-Jupiter, massing about 13 times that of our gas giant. It's orbit is some 9 AU from the star, roughly the distance of Saturn from our sun, and the orbital period is 20-some years. It rotates in about an hour. It was discovered in 2008, using the very large telescope in Chile. Beta Pictoris b has the distinction of having its own dedicated observing satellite, called PicSat. This is a 3U cubesat from the Paris Observatory, in Earth orbit. . The relatively low cost and modular design make the cubesat ideal for this type of mission.

A planet of that size has an gravitational effect on it's star, and this has been observed.

Gilese 581

Gilese is a red dwarf star in the constellation Libra, some 20 light-years from Earth. It has a mass of about 1/3 that of our sun. It has three confirmed and several unconfirmed planets. The confirmed planets are all larger than Earth, and within the habitable zone. Their orbital parameters have been determined. In addition to the planets, there is a debris disk. The first planet was discovered by the Observatory of Geneva in Switzerland in 2005, and was followed by 2 others. In 2010 the Keck Observatory in Hawaii spotted two additional planets, which have not yet been confirmed.

H209458b

Nicknamed Osiris, the exoplanet orbits the star H209458b in the constellation Pegasus. It is closer to its star than Mercury is to our Sun. It's year is 3.5 of our days. As you would expect, the surface temperature is extreme. It is known to have an atmosphere, containing hydrogen, oxygen, and carbon. It is a gas giant like Jupiter, and has a superstorm like the Great Red Spot.

WASP-3b

This exo-planet is associated with the star WASP-3, in the constellation Lyra. It is some 80 light years from Earth. It was discovered by the Southern Hemisphere-based SuperWASP observatory. The planet is classified as a gas giant, similar to Jupiter. It is closer to its sun, so is called a "hot" Jupiter.

PSR B1620-26

This was the first exoplanet confirmed to be orbiting a binary star. The star system consists of a pulsar and a white dwarf, located in the globular cluster M4. The planet is 2 ½ times the size of Jupiter. It must be a wild ride, being tugged on by 2 masses. Seven of these systems have been confirmed.

OTS-44

This is a rogue planet, not orbiting a star, or possibly a brown dwarf star, in the constellation Chameleon. It has more than 11 times the mass of Jupiter, yet is one of the lowest mass, free flying sub-stellar objects observed. It was discovered in 1998, and has been observed by the Spitzer space telescope, Herschel, and the Very Large Telescope.

Barnards Star-B

This exoplanet was just discovered as this book was being wrapped up. The star is a red dwarf, and the planet is a Super Earth, about 3.2 times larger than our home planet. It was discovered by the radial velocity method, working on 20 years of data. The planet is about 0.4 AU from its star, similar to Mercury in our system.

Alpha Centauri's planets

A special target for looking for exoplanets is Alpha Centauri, the Stat closest to out solar system, a mere 4.37 light years distant. It is actually a binary star system of Alpha Centauri A and Alpha Centauri B. There is an associated red dwarf, about 13,000 AU distant, called Alpha Centauri C, or Proxima Centauri. The European Southern Observatory discovered an Earth-sized planet orbiting Proxima Centauri in the habitable zone. This discovery came in 2016. Another small planet, smaller than Mercury, had been found in 2013, around Alpha Centauri. In 2015, the HST saw a transit event caused by a planet of about Earth's size. There may be more.

Kepler-443b

To date, the most distant confirmed found is Kepler-443b, 2,540 light years from out solar system. It seems to be in its primary's habitable zone, but has only a small chance of being rocky.

Search for Exoplanets

How do we find an exo-planet? It emits no light of its own, may reflect very little of the light from its primary, is smaller than a star, far away, and is either in the glare of its primary, or behind it, from our point of view. There are several different techniques.

Some of the super-Earth category are actually big enough to see through a telescope if conditions are right. For all the others, particularly the Earth-class, we have to use indirect methods.

The radial velocity method takes advantage of the planet's gravitation effect on the star. Variations in speed are indicated in changes to spectral lines, due to Doppler. This is accurate enough to confirm the planet's existence, and give a rough value for its mass.

The transit method uses the fact that as the exo-planet moves across the star's disk, as seen from Earth, the brightness varies slightly. This might be on the order of 10's of parts per million. If there is more than one planet, the transit timing variation method can be used to separate these out, even if they are not observable in transiting.

This method can give an estimate of the planet's size. This method has high false positive results. In addition, stars that have a brightness variation such as Red Giants cause a false positive. If an estimate of the planet's mass is available, then we know the density, and we cause postulate the composition.

The transit method can provide information on the exo-planet's atmosphere, with its star as the light source for a spectroscopic analysis.

The relativistic or Doppler beaming method uses the fact that the planet's gravity is affecting the star's brightness. This method is not very good in discovering new planets, but it does allow an estimate of the planet's mass. Bigger planets, Jupiter-sized, will cause tides in the primary, that are observable as changes in brightness.

The reflection-emission modulations method uses the fact that the planets rotate (assuming they are not in tidal lock), and variation in the thermal emission of the planet can be observed.

The class of stars called pulsars emit very regular radio waves. If the star has a planet, it will induce changes in the timing of the pulses. This is a sensitive method, allowing for detection of much smaller planets, but the trick only works when the primary is a pulsar. This implies the star has already exploded in a supernova, which is really bad for the planet. On the other hand, a pulsating variable star expands and contracts regularly, and a planet will affect this timing.

It is difficult to search for Exoplanets from the Earth's surface. It can be done with radio-telescopes. We need to search in the Infrared bands, which are mostly blocked by the atmosphere. Infrared can show us features of distant stars, where most of the visible light is scattered by space dust. Another good idea is simultaneous observations by different observatories with different point of view. This can be done with Earth-based or space-based assets, or a mix of both. This sometimes forms what is called a "virtual telescope."

A technique used to enhance our view is called gravitational micro-lensing. This takes advantage of the fact that light can be bent by a gravitational field, according to the theory of relativity. This has been observed, and exploited for exo-planet imaging. This technique requires a massive body (star) between the target and the observer. This bend the light from the distant star, and amplifies it for the observer. This was suggested by Einstein in 1915. It is somewhat of a arbitrary alignment technique, and usually does not repeat. It is effective for small exoplanets, smaller than Earth-sized.

Since exoplanets do not emit light, but only reflect the light of their primary, it is hard to image them. It is easier in the infrared spectrum, but that radiation doesn't make it through Earth's atmosphere.

What do we want to know about exoplanets, besides their existence? The mass and the radius would be nice, as that gives us the density, and hints at the composition. The temperature would tell us if liquid water could exist, and that is a precursor to life.

Astronomers first looked for planets around suns like our own. But, the first two discoveries orbited pulsars. The first discovery around a star like ours was named 51 Pegasi b. In 1998, a Jupiter-sized planet in orbit much closer to its primary, was discovered around the star Gamma Cephei.

Earth based

Earth-based optical and radio telescopes have been used in the search for exoplanets, but are hampered by the atmosphere, acting as a filter.

Radio telescopes, such as the Green Bank Observatory in West Virginia, and the University of Puerto Rico's Arecibo have been searching for patterns in the radio spectrum for decades. Greenbank has the world's largest steerable radio telescope, 100 meters in diameter. Arecibo is on the northern coast of the island, and was the largest radio telescope(at 305 meters) on the planet until recently. China built a 500 meter one. There are, world wide, more than 100 radio telescopes. They have been instrumental in discovering exoplanets.

The Green Bank facility, in West Virginia, is hosted in a small town in the mountains The radio-telescope has a 100 meter diameter. The author visited this facility when he was in grade school. The telescope I saw, a 90 meter parabolic, was built in 1962, and collapsed in 1988. It has since been replaced. The area is a designated radio-quiet zone.

The Arecibo facility is built inside a large sinkhole. It can transmit (multi-Terawatt class) as well as receive. The depression is spherical, and the receiver/transmitter is movable. This involves a

900 ton platform, suspended 500 feet above the dish. The facility can observe in a 40 degree cone around the zenith line.

The Very Large Array is a radio astronomy facility in New Mexico. It has 27 25-meter diameter dishes arranged in a Y-shape. The dishes are mounted on railroad tracks to allow adjustments. They function together as an interferometer.

Most optical telescopes are located at high altitude, to get above as much of the atmosphere as possible. Optical telescopes involved in the search for exoplanets include the Keck facility in Hawaii, and a facility high in the Andes Mountains in Chile.

Adaptive optics is a technique to improve optical telescopes by removing some of the distortion induced by travel through the Earth's atmosphere. It is better than nothing, but putting the telescope above the atmosphere is a better (yet, more complicated) solution.

A project in construction is Europe's Extremely Large Telescope. It is being built in the Atacama desert of Chile, at an elevation around 10,000 feet. It has a primary mirror of nearly 40 meters in diameter, and incorporates active optics. One of its tasks is to search for exoplanets.

The Wide Angle Search for Planets (WASP) Project is an international effort. There two robotic telescopes, one in the Northern and the other in the Southern hemisphere, scanning the sky for potential exoplanets. SuperWasp-North is in the Canary Islands. WASP-South is at the South African Astronomical Observatory. Each has eight, wide-angle cameras. Their job is to identify potential targets, that can be investigated further. Wasp is currently responsible for the discovery of some 120 exoplanets.

The Gemini Planet Imager is a instrument associated with the Gemini South Telescope in Chile, built and operated by a worldwide collaboration. It came into operation in 2014. It is specifically targeting gas giants. It found 51 Eridani b, which has the largest methane signature ever detected.

The Giant Magellan Telescope is under construction, for

completion in 2025, in Chile at an altitude of 2,500 meters. Partners include Australia, Brazil, and South Korea. It is a billion dollar (10^9) project. It will consist for seven 8.4 meter dishes for visible and near infrared. There is one central mirror surrounded by six others. This will combine to an effective 22 meter dish.

Space-based

This section discusses the ongoing search for Exoplanets with space-based resources.

HST

Earth-based telescopes have the problem that they have to look through the atmosphere. One way to solve that is to put them in orbit. This was suggested in 1923 by Hermann Oberth. The Large Space Telescope Project was funded in 1978. The spacecraft was later named after Astronomer Edwin Hubble.

The Hubble Space Telescope was placed in orbit by the Space Shuttle in 1990. It is still operating as of this writing. It is supported by the Goddard Space Flight Center, and the data goes to the Space Telescope Science Institute on the Johns Hopkins University Campus in Baltimore, MD. The idea was introduced in a paper by Lyman Spitzer in 1946, well before the "space age." Mention of the advantage of such a technique was documented as early as 1837.

The main mirror is almost 8 feet in diameter. The telescope operates in the visible spectrum, as well as near ultraviolet and near infrared. Precursor missions include the Orbiting Solar Observatory, and the Orbiting Astronautical Observatory, both from NASA's Goddard Space Flight Center. Hubble uses a Cassegrainian reflector design, from Marshall Space Flight Center, and built by Perkin-Elmer. They subcontracted the mirror to Kodak, using Corning glass. This was coated with aluminum.

After only a few weeks in orbit, it became obvious that the mirror was flawed. Corrective optics were designed, and several Shuttle flights got the orbiting facility back to its desired accuracy. At the same time, updated equipment was installed, and a failed module was replaced.

At the time, a Final Systems test was not done on HST before launch, to save money. In retrospect, this may not have been the best choice.

Hubble has a 2.4 meter, 1-ton, mirror, with corrective optics installed in orbit by astronauts from the Space Shuttle. The updates took five Shuttle Missions, and also included change-out of some of the instruments. The onboard computers were replaced, and the original tape recorder was replaced by a solid state unit All six gyroscopes were replaced, as well as the solar panels. This illustrated the versatility of repairs in space.

Hubble currently downlinks about 140 gigabytes of observation data weekly. The Hubble camera is 16 megapixels.

Like most Earth-based observatories, scientists can apply to use the HST for observations. Competition is fierce for observing time. In the period 1990-1997, the director of the StScI allowed selected amateur astronomers access to the facility for free. Thirteen amateurs were involved.

The mission was launched in 1990, repaired and updated in space by several Shuttle Missions, and continues to return good data to date. One of Hubble's notable achievements was the determination of the rate of expansion of the Universe. I did some work on this mission.

The contributions of the HST mission to our knowledge of the Universe are impressive. Hubble observations now put the estimated age of the Universe at around 14×10^9 years.

The Hubble may last another 15-20 years. It is not currently feasible to mount another repair mission, as the shuttle fleet has been decommissioned.

As this book was being written, Hubble was in safe-hold mode, due to a gyro failure. It continued to operate, although it only has 2 working gyros for pointing and positioning. The original 3 mechanical gyros, replaced on a Shuttle mission, have failed. One of the three replacement units is balky. If the third gyro behaves itself, the mission can continue. It can operate in a reduced mode, with only two. The third gyro does seem to be behaving itself, and HST went back to operational status. A robotic servicer is being developed at GSFC that might prove handy. It would be good to have HST working when JWST becomes operational, as they can work together on observations.

Dr. David Sing of JHU heads up the HST research project on exoplanets.

Wfirst

The Wide Field Infrared Survey Telescope launched in 2009. It was a NASA/GSFC mission, and observed in the infrared spectrum. It has a 2.4 meter wide field of view telescope. It has an infrared camera, and a Coronograph. That instrument is small field of view, but high contrast, in the near infrared and visible spectrum. These mission grew out of a joint NASA-DOE's Joint Dark Energy Mission. The spacecraft also assists in a census of exoplanets. It may be able to use its chronograph for direct observation of these interesting objects.

Kepler

Kepler's mission was to search for Earth-sized exoplanets. It did its job well, and is responsible for the most confirmed Exoplanets found. It was launched in 2009, and is operated by JPL. It uses a photometer to measure the brightness of stars, and that will dip as an exoplanet passes in front of the star. The brightness of some 150,000 stars is monitored. It is in an Earth-trailing heliocentric orbit.

In 2013, the second of four reaction wheels on the spacecraft failed, which disrupted operations. It is not quite a fatal failure, as the planet-hunting can still continue, under less than optimal conditions. It is currently focused on finding habitable, Earth-like planets around Red dwarf stars. The success count as of 2015 was 1,000 confirmed exoplanets discovered by Kepler., later update to 1,284 Earth-sized exoplanets in the habitable zones of their Suns.

At launch, Kepler had the highest data rate of any mission so far, because of the 95 million-pixel detector. The primary mirror is 1.4 meters in diameter, manufactured by Corning, who did the Hubble mirror. What the telescope is looking for is about a 80 parts per million decrease in brightness as the exoplanet moves in front of its star, from the telescope's point of view.

The orbiting telescope is operated and monitored by a Control Center on the Campus of the University of Colorado. Science data is downloaded once per month. A partial analysis of data is done onboard, and not all raw data is downloaded. The data management is located at the Space Telescope Science Institute on the Johns Hopkins University campus, in Baltimore, MD. Here the data is partially processed, and forwarded to NASA-Ames in California.

Kepler made major discoveries in the Gliese 581 star system. That star is a M-class red dwarf, about a third the size of our sun. It is located in the constellation Libra, about 20 light years away. The star itself has been know since around 1886.

The star has 3 confirmed planets, and 2 unconfirmed, as of this writing. The Planet Gilese 581b was discovered by the Kepler mission. Another planet, Gilese 581b is orbiting a a distance from the star that would favor liquid water, and perhaps life.

Other significant discovery's from Kepler include Kepler 10b, a rocky world, 1.4 times Earth size. Kepler, Kepler 16b, which orbits a binary star, 20e, which is smaller than Earth, 37b, which is the size of our moon. It also found 62 e & f, which seem to be water worlds, and 22b, which may be habitable.

The spacecraft was put into sleep mode, requiring none of the remaining onboard fuel. As of this writing, NASA has declared the mission over, as there is no residual fuel.

The Kepler Mission discovered over 70% of the 3,800 (at the time) confirmed exoplanets, using the transit method. This took decades of observation. In extended mission, it found some 354 more .

NASA's TESS Mission

The Transiting Exoplanet Survey Satellite is a follow-on to the Kepler mission. It uses transit photometry to spot planets orbiting distant stars. Unfortunately, only about 1% of the star systems are properly aligned to use this technique, but it still worth the effort. It was launched in 2018. The primary mission objective is to survey the brightest stars near the Earth. At its launch, about 3,800 exoplanets were know. TESS is expected to add 20,000 more. It will do a all-sky survey on about 85% of the sphere it occupies.

TESS will pass along to JWST, when that mission is launched, targets of interest. The TESS mission if lead by MIT, with funding by Google. It was built by Orbital Sciences, and launched on a Falcon-9. It is in full operation at this time. It uses four wide-field CCD cameras. TESS is operated by NASA, and the Smithsonian Astrophysical Observatory.

It's first sighting of an exoplanet was in September, 2018, as this book was in preparation.

It is planned for TESS's best targets to be passed along to JWST, when it achieves operational status, for observation.

Spitzer

When Spitzer was launched in 2003, it had a planned mission life of 2-5 year. It has been operating for 15 years now. Lyman Spitzer was an early proponent of telescopes in space.

Spitzer is an infrared telescope, which was cooled by a supply of liquid helium. When the Spitzer mission expended it's liquid helium supply in 2009, it was operated as "Warm Spitzer," in extended mission. Only two of the instruments are operable without the coolant, but that's better than nothing.

Spitzer was placed in orbit around the sun, trailing the Earth, and drifting further outward at a slow pace. Not being in orbit around the Earth, it is not exposed to the heat output of the Earth, and better observations at lower temperatures can be achieved.

The primary mirror is 85 centimeters in diameter, and was cooled to 5.5 degrees K when the liquid helium was available. This same mirror material was used on JWST.

The instruments include the IRAG, an infrared camera operating in two pairs of two chosen wavelengths with a 256 x 256 detector array. The infrared spectrometer observes at four wavelengths, using 128 x128 element detectors. The Multiband Imaging Photometer has three detector arrays.

One of the important observations that Spitzer made was of exoplanets. This was the first time an exoplanet was imaged. Usually, they are discovered by their effects on their star. The surface temperature of the exoplanet was determined, It went on to make additional contributions in the areas of formation of stars. It discovered the *double helix nebula* named because of its shape. It observed a collision between two planets orbiting a distant star, as well as a gas giant some 13,000 light years away. Spitzer and HST collaborated on the discovery.

JWST

The James Webb Space Telescope is the follow-on the HST. It uses updated technology and a new approach for the mirror, using 18 hexagonal segments, that are individually adjustable. The resulting mirror is 6.5 meters in diameter. JWST observes in long wavelength visible through the mid-infrared. The spacecraft will be placed in a halo orbit at the Earth-Sun L2 Lagrange point about 1.5 million miles from Earth. It has a large sun shield to block the

Sun's light form interfering with the observations. The project was the top pick in the 2000 Decadal Survey. Work has been going on since 1989, primarily at the Goddard Space Flight Center, As of this writing, the launch of the telescope is delayed, as the sun shield ripped in a test deployment.

JWST is a joint project of NASA, ESA, and the Canadian Space Agency. In all, fourteen countries were involving in construction of the spacecraft. John Mather, the Senior Project Scientist, had previously received the 2006 Nobel Prize in Physics.

Partially to avoid the prior HST problem with the main mirror, the JWST has a large set of small, adjustable mirrors, individually controllable with 6 actuators each. They are gold-coated beryllium. Since it observes in the infrared part of the spectrum, the detectors have to be cooled to single digits above absolute zero. A large sun shield will be used to shadow the spacecraft form the Sun, Earth, and moon. The solar wind will push the sun shade and thus the spacecraft around. But, JWST has a trim tab to counter this.

The project was a concept in the mid-1990's and construction was completed in November of 2016. It is in testing at the moment. It was originally called the Next Generation Space Telescope (NGST), but was named after James Webb, the second NASA administrator.

Where Hubble has a 2.4 meter single-piece mirror, JWST has 18 segments for a combined size of 6.5 meters. These segments are folded for launch, and deploy in orbit. It has curved secondary and tertiary mirrors. Instruments include a near infrared camera, a near infrared spectrograph, a mid-infrared instrument, and a near infrared imager with a slit-less spectrograph. The telescope is expected to be able to see extra-solar planets directly. The telescope masses 6.5 tons. It will be launched on an Ariane vehicle. It will use NASA's Deep Space Network for data transmission.

The instrument set includes a near-infrared camera (NIRCam), and spectrograph (NIRSpec). These have 32 megapixels. It uses a new type of gyro, the hemispherical resonance gyro, for attitude sensing. With essentially no moving parts, this type of gyro should last a lot longer. The spacecraft has about 60 gigabytes of onboard storage, and uses lithium-ion batteries. Onboard, it runs Java scripts for operations.

It is not designed to be serviceable. It has a projected lifetime of 5 years, with a goal of twice that. Originally estimated to cost a half-billion dollars in 1997, the current cost looks in the vicinity of 8.8 billion dollars. Actually, the spacecraft does include a docking ring for an Orion manned spacecraft (which has yet to fly), so in theory, it might be serviceable. Plan ahead.

One of the more interesting tasks for the observatory will be to assist in the search and characterization of exoplanets

It will be operated from the same facility that Hubble is, the Space Telescope Science Institute, on the campus of the Johns Hopkins University in Baltimore, MD.

Currently, Integration and Test at Northrop Grumman has been delayed into 2019, with .launch is planned for 2021. Nominal mission life is ten years. One limitation will be the station-keeping propellant supply. With current technology, JWST is not serviceable. Another possible problem is a launch slip. The 6200 kg spacecraft is scheduled to go on an Ariane-5 vehicle, which is being phased out in favor of the Ariane 6. Hopefully, HST will still be operational when JWST makes it to orbit, and the two can be used together. The ESA space telescope Herschel will be in the vicinity of L2 when JWST arrives. It was launched in 2009, and operated until 2013. Current launch date for JWST is March of 2021.

I was able to visit JWST several times, when it was in Goddard's massive clean room. Some of the components were tested in Goddard's Space Environmental Simulator (thermal-vacuum chamber). The entire telescope wouldn't fit, though. Later in the program, JWST was taken to the large chamber at Johnson Space Center, the largest in the world, developed for the Apollo program.

HabEx

NASA's Habitable Exoplanet Imaging Mission is a concept for a space-based telescope for Earth-sized exoplanets in the habitable zone. The mission is being reviewed, and may get funded for a 2020 launch to the L2 point. For Earth-sized rocky planets around main sequence stars, the telescope would analyze their atmosphere, looking for oxygen and water vapor, and, particularly bio-signature gases.

Cheops

This is a planned ESA Mission, the Characterizing ExO Planets Satellite, for launch shortly after this book will be published. It will go into a sun-synchronous orbit, with a 30 cm telescope. It will look for transiting exoplanets. It is scheduled for launch in 2019 to a sun-synchronous orbit.

The mission hopes to measure the radii of known exoplanets. For some of these, the mass is already known. These two measurements will allow calculation of the planets' density, which allows speculation of its composition. The spacecraft can sustain a downlink rate of 1.2 gigabytes per day.

ESA's PLATO Mission

The PLATO (Planetary Transits and Oscillations of Stars) mission is scheduled to be launched around 2026. It is specifically designed to look for planetary transits of stars. It is specifically looking for rocky exoplanets around yellow stars (like ours), sub-giants, and red dwarfs.

It will focus on Earth-like planets in the habitable zone around stars like out Sun where water can exist in liquid form. The specific stated goals of the mission are to:

Detect and characterize Earth-sized planets and super-Earth's in the habitable zone around solar-type stars.

Discover and characterize a large number of exo-planetary systems to study their typical architectures, and dependencies on the properties of their host stars and the environment.

Measure stellar oscillations to study the internal structure of stars and how it evolves with age.

Identify good targets for spectroscopic measurements to investigate exoplanet atmospheres.

COROT

The Convection, Rotation et Transits planétaires is a French mission that operated from 2006 through 2013. It was looking for large, terrestial sized planets with short orbital periods. It also was conducting research in the seismology of stars. It discovered the first exo-planet known to have a rocky or metalic composition. The mission used the transiting approach to find exoplanets. Its first discovery was in 2007, and it went on to discover some 30 more exoplanets. It operated until a computer problem caused the loss of data from the instrument. The spacecraft was decomissioned, and left to reenter the atmosphere.

ESA's GAIA Mission

The GAIA mission launched in 2013, with ambitious goals. It was to construct the largest and most precise catalog of space objects, in three dimensions. It will monitor these objects over a period of years. It is expected to catalog over a billion objects, although this is only about 1% of the population of the Milky Way. Necessarily, some of the observed stars will have associated planets. One goal is to determine the orbits and inclination of extra-solar planets, as well as their masses.

European Ariel Mission

This mission will launch in 2028, and will observe 1,000 selected exoplanets, and survey the chemical composition of their atmospheres.

Canada's MOST Satellite

Canada's first space telescope project specifically focuses on variations in the observed light from stars. There are many causes for this, one of which is a planet passing through the field of view. Once a target is selected, it will stay focused on that target for up to two months. It is conducting research in astro-seismology, and exo-planet transits.

Due to government funding issues, the spacecraft and operations were acquired by a private company, Microsat Systems Canada Inc. (MSCI).

Europa Clipper Mission

This NASA pending mission has a goal of exploring our solar system's Ocean worlds to get a better understanding of them, and their ex0-counterparts. The mission, to be launched around 2025, will study Jupiter's moon Europa, in a follow-on to the earlier Galileo spacecraft. Due to the intensity of the Jupiter radiation belt, it will not orbit Europa, but rather fly an elliptical orbit around Jupiter, avoiding the belts as possible, and do multiple fly-bys of Europa.

ESA will launch its Jupiter Icy Moons Explorer in 2022, and conduct several fly-by's of Europa as well, on its way to Callisto and Ganymede. Characterizing the planets in our system will show us what to look for in other systems.

Interstellar probe

When are we going to send out an inter-stellar mission? Well, actually we did that in 1977. The two Voyager spacecraft were launched that year, following Pioneer 10 and 11. Recently, New Horizons, completing its mission at Pluto, headed out to join the others in Interstellar space. Voyager I reached interstellar space in 2013. What is interstellar space? It's the point where our Sun's influence begins to wane, and the spacecraft has sufficient velocity to continue indefinitely. The boundary is around 122 AU. Pioneer 10 and 11 are inoperable. I worked on the Voyager missions, which continue to send back good data.

Interstellar Probe was a 1999 JPL concept sending a spacecraft to 200 AU in 15 years. It would have used a solar sail out to 5 A.U. Another concept used an ion engine powered by RTG's.

If we launched a probe today, it could reach our closest neighbor Alpha Centauri in 40,000 years. We would really like to get some data in our lifetimes. NASA has discussed an interstellar mission for 2069.

Afterword

The Universe is so large, and has so many interesting places, while being relatively unexplored, will wee discover other life, or be discovered by it. Are we ready?

Glossary

3U – 3 Units of of 10 x 10 x10 Centimeters.
Adaptive optics – adjusting a mirror to adjust for distortion,
Albedo – ratio of irradiance reflected to irradiance received.
ALMA - Atacama Large Millimeter Array, Chilean Andes.
AMFE – a message from Earth.
AMNH – American Museum of Natural History, New York City
Angstrom – measure of length, 0.1 nanometer.
APL – Applied Physics Lab of the Johns Hopkins University.
Arc-minute – 1/60 of a degree.
Arc-second – 1/3660 of a degree.
Areoid – reference zero-elevation surface.
Ariel - Atmospheric remote sensing infrared exoplanet large survey.
ASIN – Amazon Standard Inventory Number.
Astrometry – measurements of positions and movements of objects in space.
Asteroseismology – study of oscillations in stars.
ATA – Allen telescope array.
AU – astronomical unit, mean distance from the Earth to the Sun, 93,000,000 miles.
BIS – British Interplanetary Society
Blazar – active galactic nucleus with a relativistic jet.
BPP – breakthrough Propulsion Physics program.
Brown Dwarf – failed star. Bigger than a planet.
ccd – charge coupled device, a sensor.
Centaur – a minor planet in an unstable orbit, behaving like an asteroid or comet.
Cepheid variable stars – star that has regular pulsations.
Ceres – dwarf planet, largest object in asteroid belt.
Cheops – Characterising Exoplanets Satellite, ESA.
Chthonian planet – orbits close to its star,
Circum-binary planet – planet orbiting two stars
Comet – a solar system object consisting of ice, dust, and gas, in highly eccentric orbit.

CoRoT – (French) Convection, Rotation et Transits planétaires.
Cosmic ray – high energy radiation, from outside the solar system.
CSA – Canadian Space Agency.
CSR – Center for SETI Research.
Cryovolcanism – ice volcano.
Dark energy – hypothetical form of energy that explains why the Universe is expanding.
Dark Matter – existence postulated. Might account for 85% of the matter in the known universe.
DISH – parabolic antenna, usually steerable.
DNA – molecule carrying genetic information in all known life.
Dwarf planet – planet below a certain size.
Dwarf star – small star, much smaller than our Sun. Comes in white, red, blue and black variations.
Eccentric Jupiter – large planet orbiting a star in an eccentric orbit.
Eclipsing binary – two stars n a common orbit pass in front of each other, from our point of view.
ELT – Extremely Large Telescope, in construction in Chile.
Enceladus – ocean moon of Saturn.
ESA – European Space Agency
ESI – Earth Similarity index
HARPS - High Accuracy Radial velocity Planet Searcher.
Exoplanets – planets from outside our solar system
Extremophiles – organism that lives in an extreme environment, unsuited to conditions favorable to life as we know it.
ev – electron volt, unit of energy.
ExoMoon – moon orbiting an Exoplanet.
Exoplanet – a planet outside of out solar system.
Exoplanetology – study of exoplanets.
FAST - Five hundred meter Aperture Spherical Telescope (China).
FRB – Fast radio burst.
Galaxy – a loosely coupled collection of 10^8 to 10^{14} stars.
Gas dwarf - A low-mass planet composed primarily of hydrogen and helium.
Gas giant - A massive planet composed primarily of hydrogen and helium.
GBT – Green Bank Telescope.

Globular cluster - a spherical collection of stars that orbits a galactic core.
Goldilocks Zone – area that is neither too hot nor too cold; from a Fairy Tale.
GPI – Gemini Planet Imager.
GRB - Gamma Ray Bursts.
Habitable zone – distance of a planet from its star that allows liquid water on the surface
HabEx - Habitable Exoplanet Imaging Mission.
HARPS - High Accuracy Radial velocity Planet Searcher.
Heliosphere – a volume of space dominated by the Sun. In our case, out beyond Pluto.
Hipparcos - High precision parallax collecting satellite.
Hot Jupiter – planet close to its star, and having a mass similar to Jupiter.
Hot Neptune – planet close to its star, and about the mass of Uranus or Neptune.
HST – Hubble Space Telescope.
Hubble – Space Telescope named after Edwin Hubble.
Hubble Constant – rate of expansion of the universe.
IAC - Instituto de Astrofísica de Canarias, observatory in Canary Islands.
IAU - International Astronomical Union.
Ice giant – large planet consisting of ices of various substances – in our system, Uranus and Neptune.
Infrared – electromagnetic radiation with wavelength longer than visible light.
ING – Isaac Newton Group of (3) Telescopes.
Interferometer – array of telescopes that work together.
Iron planet - A type of planet that consists primarily of an iron-rich core with little or no mantle.
ISBN – International Standard Book Number.
ISM – interstellar medium
JPL – NASA's Jet Propulsion Lab, Pasadena, California.
JWST – James Webb Space Telescope.
KBO – Kuiper Belt objects.
KOI – Kepler Objects of Interest.

L2 – LaGrange point (null in the gravity field of 3 bodies. L2 is located on the far side of the smaller mass.
LGM – little green men.
Light-year - distance light can travel in a year – 9,460,730,472,580,800 meters.
LLNL - Lawrence Livermore National Lab
Main Sequence – band of stars on plot of color vs brightness.
MCC – Mission Control Center.
MSCI - Microsat Systems Canada Inc.
MEV – million electron volts, a measure of energy.
Moon – smaller astronomical body in orbit around a planet.
MoonMoon – small body orbiting the moon of a planet.
MOST - Micro-variability and Oscillations of Stars (Canada)
NASA – (U. S.) National Aeronautics and Space Administration.
Nebula – interstellar cloud of dust and gasses.
NGC - New General Catalogue of Nebulae and Clusters of Stars.
NIROSETI - Near-InfraRed Optical Search for Extraterrestrial Intelligence.
NOM – natural organic matter.
NRAO - National Radio Astronomy Observatory
NSSDC -National Space Science Data Center.
NWO - Nederlandse Organisatie voor Wetenschappelijk Onderzoe.
Ocean planets – all of the surface is covered with water.
Organic – pertaining to a living organism.
Oort cloud – icy small planets at 2,000 to 200,000 AU from the Sun.
Oumuamua – Asteroid not from our solar system.
Outer Planet – in our solar system, planets beyond the asteroid belt.
OWEP – Ocean Worlds Exploration Program.
Parsec – parallel second of arc, unit of length, about 3.26 light years.
Planet – a body orbiting a star.
PLATO - PLAnetary Transits and Oscillations of stars.
PPD – proto-planetary disk – rotating circumstellar disk of gas and dust, around a young star.
Puffy planet - Gas giant with a large radius and very low density.

Pulsar – highly magnetized rotating neutron star of white dwarf.
Pyroclastic flow – hot gas and volcanic matter from a volcano.
Pulsating variable star - swells and shrinks periodically.
Radial velocity – perturbations in a star, due to an orbiting planet.
Red Dwarf – small and cool star; most common in the Milky Way.
RGO – Royal Greenwich Observatory (U.K.)
Ring system – a disk of solid material around a planet.
RNA - Ribonucleic acid, biological molecule.
Roche limit - the distance in which a celestial body, held together by its own gravity, will disintegrate due to tidal forces of its primary.
Rocky – consisting of silicates and metals.
Rogue planet – orbits the galactic center, not a star.
Ross-128b – confirmed Earth-sized Exoplanet.
RTG – radioisotope thermoelectric generator.
SAO - Smithsonian Astrophysical Observatory.
Scintillation – variations in apparent brightness.
SDO – scattered disk object.
SERENDIP - Search for Extraterrestrial Radio Emissions from Nearby Developed Intelligent Populations.
SETI – Search for extra-terrestrial intelligence.
SFL – Space Flight Laboratory, University of Toronto, Canada.
SIMPS - Supplemental IRAS Minor Planet Survey.
SKA – square kilometer array (South Africa).
Snow line/ice line – point beyond which there is no liquid water.
Solar flare – a sudden rapid emission of electrons, ions, and atoms from a star.
Solar System – A star and its associated planets and such.
Solar wind – stream of charged particles emitted from a star's upper atmosphere.
Spectroscopy – analyzing light by breaking it down into its individual wavelengths.
SPT – South Pole (radio) Telescope.
SSO – Swiss Space Office.
Star - luminous ball of plasma.
STFC – (U.K.) Science and Technology Facilities Council.
STS – Space Transportation System (Shuttle).

StScI – Space Telescope Science Institute (JHU)
Super Earth – extra-solar planet, larger than our Earth.
Supernova – very large explosion of an end-of-life star.
TBD – to be determined.
Tera-watt – 10^{12}
Terrestrial planet – Earth-like, composed primarily of carbonaceous or silicate rocks or metals.
TESS – Transiting Exoplanet Survey Satellite.
Tholin – organic compound formed by UV radiation from carbon compounds.
Tidal lock – where the same side of a object always faces the primary it is orbiting.
TMT – thirty meter telescope.
TNO – trans-Neptunium object.
Transit – when an exo-planet crossed the face of a star, as seen from Earth,
Trappist - Transiting Planets and Planetesimals Small Telescope (Chile).
Trojan – minor planet that shares an orbit with one of the larger planets.
USGS – United States Geological Survey.
UV – ultraviolet, 19 nm to 400 nm wavelength.
Virtual Telescope – using multiple telescopes simultaneously for one observation.
VLA – very large array.
VLBA – Very large baseline array.
WASP – Wide Angle Search for Planets.
Waterworld – exo-planet completely covered by ocean. Postulated, not yet observed.
WFIrST - Wide Field Infrared Survey Telescope.
White dwarf – remnant of a collapsed star. No fusion activity.
X-ray - 0.1 to 10 nanometer wavelength.
X-ray binary (star) – binary star, emitting x-rays.
YSO – young stellar objects.

Bibliography

Anderson, Mathew *Habitable Exoplanets: Red Dwarf Systems Like TRAPPIST-1* (OCS Book 3), 2018, ASIN-B07C69B5VJ.

Anderson, Mathew *Is Anybody Out There?: An Expanded Excerpt From Our Cosmic Story* (OCS Book 2), 2017 ASIN-B071VBCX54.

Armitage, Philip J. *Astrophysics of Planet Formation*, 1st Edition, 2009, ISBN-0521887453.

Billings, Lee *Five Billion Years of Solitude: The Search for Life Among the Stars*, 2013, ISBN-1617230065.

Cockell, Charles S. *Astrobiology: Understanding Life in the Universe,* 1st Edition, ISBN-1118913337.

Culp, Jennifer *How We Find Other Earths: Technology and Strategies to Detect Planets Similar to Ours (Search for Other Earths)*, 2016, ISBN-1499462921.

Deroo, Pieter *Spectroscopic Characterization of Extra-solar Planet Atmospheres,* NASA, 2012, ASIN-ASIN-B01FKLKX6Y.

Gargaud, Muriel; Martin, Hervé *Young Sun, Early Earth and the Origins of Life: Lessons for Astrobiology*, 2013, ISBN-3642225519.

Goldsmith, Donald *Exoplanets: Hidden Worlds and the Quest for Extraterrestrial Life,* Harvard University Press, 2018, ASIN-B07DGJKF8N.

Hart, Chris; Vernekar, Shubham *After Earth : The Search for Habitable Exoplanets: The Search for Habitable Exoplanets*, 2017, ASIN-B076YBDHZC.

Haswell, Carole A. *Transiting Exoplanets*, Cambridge University Press, 2010, ISBN-0521139384.

Heng, Kevin *Exoplanetary Atmospheres: Theoretical Concepts and Foundations* (Princeton Series in Astrophysics), 2017, ISBN-0691166978.

Johnson, John Asher *How Do You Find and Exoplanet?*, Princeton University Press, ISBN-0691156816.

Kenney, Karen Latchana *Exoplanets: Worlds beyond Our Solar System,* 2017, ASIN-B01N2RSA5H.

Kisak, Paul F. (ed) *Abiogenesis: Natural Processes for the Origin of Life*, 2016, ISBN-1537072900.

Kitchin, C. R. *Exoplanets: Finding, Exploring, and Understanding Alien Worlds*, 2012, ISBN-1461406439.

Linde, Peter *The Hunt for Alien Life: A Wider Perspective* (Astronomers' Universe), 2016, ISBN-3319241168.

Lissauerby, Jack J.; de Pater, Imke, *Fundamental Planetary Science,* 1st Edition, Wiley Blackwell, 2015, ISBN-0521853303.

Mammana, Dennis; McCarthy, Donald *Other Suns. Other Worlds?: The Search for Extra Solar Planetary Systems*, 1996, ISBN-0312140215.

Micheli, Marco et al, "Non-gravitational acceleration in the trajectory of 1I/2017 U1 ('Oumuamua)," 27 June 2018, avail: https://www.nature.com/articles/s41586-018-0254-4.

Ohlmeier, Astrid Z. *Planets: The Solar System & Extra-Solar Planets*, NASA, 2014, ISBN-1494923464.

Perryman, Michael *The Exoplanet Handbook*, 1st Edition, 2014, ISBN-0521765595.

Pont, Frederic J. *Alien Skies: Planetary Atmospheres from Earth to Exoplanets*, 2014, ISBN-1461485533.

Scharf, Caleb *The Copernicus Complex: Our Cosmic Significance in a Universe of Planets and Probabilities*, 2014, ISBN-0374129215.

Seager, Sara *Exoplanet Atmospheres: Physical Processes,* 2010, Princeton University Press, ISBN-0691146454.

Sengupta, Sujan *Worlds Beyond Our Own: The Search for Habitable Planets*, 2014, ISBN-3319098937.

J.M. Shull, J. M., Thronson Jr. Harley A. *The Search for Extra-Solar Terrestrial Planets: Techniques and Technology*: Proceedings of a Conference held in Boulder, Colorado, May 14–17, 1995, ISBN-079234474X.

Smith, Eric *The Origin and Nature of Life on Earth: The Emergence of the Fourth Geosphere*, 1st Edition, 2016, Cambridge University Press, ISBN-1107121884.

Steves, Bonnie, Hendry, Martin *Extra-Solar Planets: The Detection, Formation, Evolution and Dynamics of Planetary Systems*, 2010, ISBN-1420083449.

Stakem, Patrick H. *Exploration of the Gas Giants, Space Missions to Jupiter, Saturn, Uranus, and Neptune*, PRRB Publishing, 2018, ISBN-9781717814500.

Summers, Michael E. *Exoplanets: Diamond Worlds, Super Earths, Pulsar Planets, and the New Search for Life beyond Our Solar System,* 2017, Smithsonian, ISBN-1588345947.

Tasker, Elizabeth *The Planet Factory: Exoplanets and the Search for a Second Earth,* 2017, ISBN-1472917723.

U. S. Government, *Complete Guide to the Kepler Space Telescope Mission and the Search for Habitable Planets and Earth-like Exoplanets - Planet Detection Strategies, Mission History and Accomplishments,* ASIN-B00CZE0E28.

Vainio, Olli-Pekka, *Cosmology in Theological Perspective: Understanding Our Place in the Universe,* 2018, ASIN-B07933B1S8.

Yaqoob, Tahir *Exoplanets and Alien Solar Systems,* 2011, ASIN-B005WQ0E6C.

Resources

- https://exoplanets.nasa.gov/
- https://www.nasa.gov/feature/jpl/what-in-the-world-is-an-exoplanet
- https://exoplanetarchive.ipac.caltech.edu/
- https://www.swri.org/press-release/evidence-complex-organic-molecules-enceladus
- www.planetary.org
- Space Telescope Science Institute public outreach, http://outreachoffice.stsci.edu/
- https://www.space.com/17738-exoplanets.html
- https://www.nasa.gov/tess-transiting-exoplanet-survey-satellite
- https://www.lpi.usra.edu/opag/meetings/feb2016/presentations/day-1/08-Roadmap-Ocean-Worlds-McEwen.pdf
- https://spacenews.com/ocean-worlds-discoveries-build-case-for-new-missions/
- www.planethunters.org
- http://phl.upr.edu/projects/habitable-exoplanets-catalog/exoplanet-resources
- SETI Institute – seti.org
- https://gizmodo.com/tag/exoplanets
- https://en.wikipedia.org/wiki/List_of_astronomy_acronyms
- Sagan, et all, "A search for life on Earth from the Galileo spacecraft, " avail:
- http://www.pw.physics.uiowa.edu/~dag/publications/1993_ASearchForLifeOnEarthFromTheGalileoSpacecraft_NATURE.pdf
- Extrasolar Planets Encyclopaedia - http://exoplanet.eu
- https://exoplanetarchive.ipac.caltech.edu
- Tau Zero Foundation - https://tauzero.aero/
- Exosolar Planets Encyclopedia - http://exoplanet.eu/
- https://exoplanets.nasa.gov/exep/

- Exoplanets-101, avail: https://exoplanets.nasa.gov/the-search-for-life/exoplanets-101/
- https://archive.is/20130929121408/http://sci.esa.int/gaia/40577-extra-solar-planets/
- http://www.planetary.org/explore/space-topics/exoplanets/
- http://www.planetary.org/explore/projects/exoplanets/alpha-centauri-planet-search.html
- www.Exoplanets.org
- https://www.space.com/40699-exoplanets-virtual-tour-nasa-tool.html
- http://exoplanet.eu/
- Planetary Society – www.planetary.org
- SETI – www.seti.ogr
- http://phl.upr.edu/projects/habitable-exoplanets-catalog
- https://exoplanetarchive.ipac.caltech.edu/
- wikipedia, various.

If you enjoyed this book, you might also be interested in some of these.

Stakem, Patrick H. *16-bit Microprocessors, History and Architecture*, 2013 PRRB Publishing, ISBN-1520210922.

Stakem, Patrick H. *4- and 8-bit Microprocessors, Architecture and History*, 2013, PRRB Publishing, ISBN-152021572X,

Stakem, Patrick H. *Apollo's Computers*, 2014, PRRB Publishing, ISBN-1520215800.

Stakem, Patrick H. *The Architecture and Applications of the ARM Microprocessors*, 2013, PRRB Publishing, ISBN-1520215843.

Stakem, Patrick H. *Earth Rovers: for Exploration and Environmental Monitoring*, 2014, PRRB Publishing, ISBN-152021586X.

Stakem, Patrick H. *Embedded Computer Systems, Volume 1, Introduction and Architecture*, 2013, PRRB Publishing, ISBN-1520215959.

Stakem, Patrick H. *The History of Spacecraft Computers from the V-2 to the Space Station*, 2013, PRRB Publishing, ISBN-1520216181.

Stakem, Patrick H. *Floating Point Computation*, 2013, PRRB Publishing, ISBN-152021619X.

Stakem, Patrick H. *Architecture of Massively Parallel Microprocessor Systems*, 2011, PRRB Publishing, ISBN-1520250061.

Stakem, Patrick H. *Multicore Computer Architecture*, 2014, PRRB

Publishing, ISBN-1520241372.

Stakem, Patrick H. *Personal Robots*, 2014, PRRB Publishing, ISBN-1520216254.

Stakem, Patrick H. *RISC Microprocessors, History and Overview,* 2013, PRRB Publishing, ISBN-1520216289.

Stakem, Patrick H. *Robots and Telerobots in Space Application*s, 2011, PRRB Publishing, ISBN-1520210361.

Stakem, Patrick H. *The Saturn Rocket and the Pegasus Missions, 1965,* 2013, PRRB Publishing, ISBN-1520209916.

Stakem, Patrick H. *Visiting the NASA Centers, and Locations of Historic Rockets & Spacecraft,* 2017, PRRB Publishing, ISBN-1549651205.

Stakem, Patrick H. *Microprocessors in Space*, 2011, PRRB Publishing, ISBN-1520216343.

Stakem, Patrick H. Computer *Virtualization and the Cloud*, 2013, PRRB Publishing, ISBN-152021636X.

Stakem, Patrick H. *What's the Worst That Could Happen? Bad Assumptions, Ignorance, Failures and Screw-ups in Engineering Projects, 2014,* PRRB Publishing, ISBN-1520207166.

Stakem, Patrick H. *Computer Architecture & Programming of the Intel x86 Family, 2013,* PRRB Publishing, ISBN-1520263724.

Stakem, Patrick H. *The Hardware and Software Architecture of the Transputer*, 2011,PRRB Publishing, ISBN-152020681X.

Stakem, Patrick H. *Mainframes, Computing on Big Iron*, 2015, PRRB Publishing, ISBN- 1520216459.

Stakem, Patrick H. *Spacecraft Control Centers*, 2015, PRRB Publishing, ISBN-1520200617.

Stakem, Patrick H. *Embedded in Space,* 2015, PRRB Publishing, ISBN-1520215916.

Stakem, Patrick H. *A Practitioner's Guide to RISC Microprocessor Architecture*, Wiley-Interscience, 1996, ISBN-0471130184.

Stakem, Patrick H. *Cubesat Engineeering*, PRRB Publishing, 2017, ISBN-1520754019.

Stakem, Patrick H. *Cubesat Operations*, PRRB Publishing, 2017, ISBN-152076717X.

Stakem, Patrick H. *Interplanetary Cubesats*, PRRB Publishing, 2017, ISBN-1520766173 .

Stakem, Patrick H. Cubesat Constellations, Clusters, and Swarms, Stakem, PRRB Publishing, 2017, ISBN-1520767544.

Stakem, Patrick H. *Graphics Processing Units, an overview*, 2017, PRRB Publishing, ISBN-1520879695.

Stakem, Patrick H. *Intel Embedded and the Arduino-101, 2017,* PRRB Publishing, ISBN-1520879296.

Stakem, Patrick H. *Orbital Debris, the problem and the mitigation*, 2018, PRRB Publishing, ISBN-*1980466483*.

Stakem, Patrick H. *Manufacturing in Space*, 2018, PRRB Publishing, ISBN-1977076041.

Stakem, Patrick H. *NASA's Ships and Planes*, 2018, PRRB Publishing, ISBN-1977076823.

Stakem, Patrick H. *Space Tourism*, 2018, PRRB Publishing, ISBN-

1977073506.

Stakem, Patrick H. *STEM – Data Storage and Communications*, 2018, PRRB Publishing, ISBN-1977073115.

Stakem, Patrick H. *In-Space Robotic Repair and Servicing*, 2018, PRRB Publishing, ISBN-1980478236.

Stakem, Patrick H. *Introducing Weather in the pre-K to 12 Curricula, A Resource Guide for Educators*, 2017, PRRB Publishing, ISBN-1980638241.

Stakem, Patrick H. *Introducing Astronomy in the pre-K to 12 Curricula, A Resource Guide for Educators*, 2017, PRRB Publishing, ISBN-198104065X.
Also available in a Brazilian Portuguese edition, ISBN-1983106127.

Stakem, Patrick H. *Deep Space Gateways, the Moon and Beyond*, 2017, PRRB Publishing, ISBN-1973465701.

Stakem, Patrick H. *Exploration of the Gas Giants, Space Missions to Jupiter, Saturn, Uranus, and Neptune*, PRRB Publishing, 2018, ISBN-9781717814500.

Stakem, Patrick H. *Crewed Spacecraft*, 2017, PRRB Publishing, ISBN-1549992406.

Stakem, Patrick H. *Rocketplanes to Space*, 2017, PRRB Publishing, ISBN-1549992589.

Stakem, Patrick H. *Crewed Space Stations,* 2017, PRRB Publishing, ISBN-1549992228.

Stakem, Patrick H. *Enviro-bots for STEM: Using Robotics in the pre-K to 12 Curricula, A Resource Guide for Educators,* 2017, PRRB Publishing, ISBN-1549656619.

Stakem, Patrick H. *STEM-Sat, Using Cubesats in the pre-K to 12 Curricula, A Resource Guide for Educators*, 2017, ISBN-1549656376.

Stakem, Patrick H. *Lunar Orbital Platform-Gateway*, 2018, PRRB Publishing, ISBN-1980498628.

Stakem, Patrick H. *Embedded GPU's*, 2018, PRRB Publishing, ISBN- 1980476497.

Stakem, Patrick H. *Mobile Cloud Robotics*, 2018, PRRB Publishing, ISBN- 1980488088.

Stakem, Patrick H. *Extreme Environment Embedded Systems*, 2017, PRRB Publishing, ISBN-1520215967.

Stakem, Patrick H. *What's the Worst, Volume-2*, 2018, ISBN-1981005579.

Stakem, Patrick H., *Spaceports*, 2018, ISBN-1981022287.

Stakem, Patrick H., *Space Launch Vehicles*, 2018, ISBN-1983071773.

Stakem, Patrick H. *Mars*, 2018, ISBN-1983116902.

Stakem, Patrick H. *X-86, 40th Anniversary ed*, 2018, ISBN-1983189405.

Stakem, Patrick H. *Lunar Orbital Platform-Gateway*, 2018, PRRB Publishing, ISBN-1980498628.

Stakem, Patrick H. *Space Weather*, 2018, ISBN-1723904023.

Stakem, Patrick H. *STEM-Engineering Process*, 2017, ISBN-1983196517.

Stakem, Patrick H. *Space Telescopes,* 2018, PRRB Publishing, ISBN-1728728568.

Stakem, Patrick H. *Exoplanets*, 2018, PRRB Publishing, ISBN-9781731385055.

Stakem, Patrick H. *Planetary Defense*, 2018, PRRB Publishing, ISBN-9781731001207.

Patrick H. Stakem *Exploration of the Asteroid Belt*, 2018, PRRB Publishing, ISBN-1731049846.

Patrick H. Stakem *Terraforming*, 2018, PRRB Publishing, ISBN-1790308100.

Patrick H. Stakem, *Martian Railroad,* 2019, PRRB Publishing, ISBN-1794488243.

Patrick H. Stakem, *Exoplanets,* 2019, PRRB Publishing, ISBN-1731385056.

Patrick H. Stakem, *Exploiting the Moon,* 2019, PRRB Publishing, ISBN-1091057850.

Patrick H. Stakem, *RISC-V, an Open Source Solution for Space Flight Computers,* 2019, PRRB Publishing, ISBN-1796434388.

Patrick H. Stakem, *Arm in Space*, 2019, PRRB Publishing, ISBN-9781099789137.

Patrick H. Stakem, *Extraterrestrial Life*, 2019, PRRB Publishing, ISBN-978-1072072188.

Patrick H. Stakem, *Space Command*, 2019, PRRB Publishing, ISBN-978-1693005398.

www.ingramcontent.com/pod-product-compliance
Lightning Source LLC
Chambersburg PA
CBHW030511220526
45464CB00006B/2756